YOUR KNOWLEDGE HAS VALUE

- We will publish your bachelor's and
  master's thesis, essays and papers

- Your own eBook and book -
  sold worldwide in all relevant shops

- Earn money with each sale

Upload your text at www.GRIN.com
and publish for free

Bibliographic information published by the German National Library:

The German National Library lists this publication in the National Bibliography; detailed bibliographic data are available on the Internet at http://dnb.dnb.de .

Imprint:

Copyright © 2014 GRIN Verlag
Print and binding: Books on Demand GmbH, Norderstedt Germany
ISBN: 9783668719996

This book at GRIN:

https://www.grin.com/document/424259

Arvind Kumar

Augmenting medicinal plant research through in vitro and in silico approach

GRIN Verlag

1 ABSTRACT

Medicinal plants are the great natural resources but due to lack of knowledge, arbitrary use and lack of conservation measures many important medicinal plant species are becoming extinct, endangered and threatened. In the present research, micropropagation studies were carried out on ten medicinal plants, showed best respond in **seven** medicinal plants among them. Callus cultures and shoot cultures were successfully initiated on basal MS media supplemented with different plant growth regulators (2,4-D, NAA, IAA and BAP) of various concentrations. These can be beneficial aspect in future as for exporting or making available the plants to farmers of superior genotype .This would also facilitate in metabolite extraction and supply in drug industries.

An online static database in the form of a web site was also created named as **"MedDBase"** (available online at https://sites.google.com/site/meddbase/). This database facilitates the medicinal as well as tissue culture information of all Indian medicinal plants. The list contained Indian medicinal plants more than 5000, is retrieved from the NMPB site and arranged in Alphabetical order. This online database will help the scientific community to keep themselves updated with the research and development work being carried out for a particular medicinal plant. This will also help in reducing duplication/ repetition of same work.

2 INTRODUCTION

India is one of the 12-mega biodiversity centres having about 10% of the world's biodiversity wealth, distributed across 16 agro climatic zones. Out of 17,000 species of higher plants reported to occur within India, 7500 of which are for medicinal uses (Lichterman 2004). This proportion of the medicinal plant is the highest known in any other country against the existing flora of that country (Lichterman 2004; Tapsell *et al.*, 2006). Ayurveda; the oldest medicinal system in the Indian subcontinent has alone reported approximately 2000 medicinal plant species, followed by the Siddha and Unani medical system. Currently 25% of drugs are derived from plants and many others are synthetic analogue built on prototype compounds isolated from plant species in modern pharmacopoeia.

There are several stakeholders in the medicinal plants sector, right from herb collectors and growers to manufacturers to consumers. More than 700,000 practitioners of Ayurveda, Siddha, Unani, Yoga, Naturopathy, and Homeopathy are registered in the Indian system of medicine and sizeable numbers of practioners are not registered. There are 9493 manufacturing units, 22,635 dispensaries and 1355 hospitals of Indian system of medicine (Fabricant and Farnsworth 2001). Approximately 800 species of medicinal plants are in active trade and still there is a gap of 40,000 metric tonnes in the demand and supply of medicinal plants. The major source of medicinal plants is the forested area and about 90% medicinal plants is collected from the wild, which generates about 40 million man-days (Billing and Sherman 1998).

Medicinal plants are not only a major resource base for the traditional medicine & herbal industry but also provide livelihood and health security to a large segment of Indian population. The domestic trade of the AYUSH industry is of the order of Rs. 80 to 90 billion (1US$ = Rs.50). The Indian medicinal plants and their products also account of exports in the range of Rs. 10 billion.

There is global resurgence in traditional and alternative health care systems resulting in world herbal trade, which stands at US$ 120 billion and is expected to reach US$ 7 trillion by 2050. Indian share in the world trade, at present, however, is quite low.

Because "over 50% of prescription drugs are derived from chemicals first identified in plants (Li and Vederas 2009), a 2008 report from the Botanic Gardens Conservation International (representing botanic gardens in 120 countries) warned that "cures for things such as cancer and HIV may become 'extinct before they are ever found'." They identified 400 medicinal plants at risk of extinction from over-collection and deforestation, threatening the discovery of future cures for disease. These included Yew trees (the bark is used for the cancer drug paclitaxel); Hoodia (from Namibia, a potential source of weight loss drugs); half of Magnolias (used as Chinese medicine for 5,000 years to fight cancer, dementia and heart disease); and Autumn crocus (for gout). Their report said, "five billion people still rely on traditional plant-based medicine as their primary form of health care.

The National Medicinal Plants Board (NMPB) set-up in November 2000 by the Government of India has the primary mandate of coordinating all matters relating to medicinal plants and support policies and programmes for growth of trade, export, conservation and cultivation.

With a view to strengthen the medicinal plant sector all over the country as well as to conserve the wild stock, the NMPB was set up by the Government of India on 24 November 2000. The prime objective of setting up the Board was to establish an agency, which would be responsible for coordination of all matters with respect to the medicinal plant sector, including drawing up policies and strategies for in situ conservation, cultivation, harvesting, marketing, processing, drug development etc.

The importance of plants has been realized and well documented by scholars since ancient period. Apart from the innumerable social benefits, much emphasis has been accorded to the plants of medicinal value. Majority of the population in developing countries depend on traditional system of medicine for their primary health care. Due to this increasing trend towards use of alternative system of medicine, natural medicinal plant resource in this world is under enormous pressure. Several Institution/ Organization/ Universities/ Pharmaceutical Industries across the world have been engaged in research and documentation of various aspect of these medicinal plants to frame a strategy for their conservation and sustainable utilization. In this way numerous research papers on each medicinal plant covering broad subject areas like Botany, Chemistry, Pharmacology, Pharmacy etc. are available. However, no comprehensive document is available, where all the information about a particular medicinal plant can be found. This led our team to conceive the idea of developing a database in which, all the available published information on selected medicinal plants, covering every subject area can be accessed at one place. This will help the scientific community to keep themselves updated with the research and development work being carried out for a particular medicinal plant. This database will also be a source of useful information for students, teachers, practitioners and all those who are involved in their cultivation propagation etc. This will also help in eliminating duplication/ repetition of same work.

2.1 OBJECTIVES

The medicinal plants sector is largely unregulated and not studied properly even at the national level. Continuous exploitation of several medicinal plant species from the wild and substantial loss of their habitat during the past 15 years have resulted in the population decline of many high value medicinal plant species over the years. This led us to carry out research out tissue culture technology for the large scale and sustainable production of quality planting material of elite medicinal plants and to develop a database of Indian medicinal plants with the following objectives:

1. Establishment of tissue culture protocol for seven medicinal plants.
2. Standardizing the physicochemical conditions for large-scale *in vitro* multiplication.
3. Standardization of various media types and growth regulators.
4. Studying the competency of *in vitro* plants to sustain the subculture durations.
5. Collection of data of medicinal plants for creating a database (an online webpage).
6. Compiling and designing of database.
7. Publishing the database online.

2.2 PLANT SPECIES SELECTED FOR PRESENT RESEARCH

2.2.1 *Aloe barbadensis* MILL.

Aloe vera is an important medicinal plant that belongs to the family Liliaceae. The flowers are hermaphrodite; plant prefers light (sunny weather), requires well-drained soil and can grow in nutritionally poor soil. Pharmaceutical and cosmetic industry has great demand in *Aloe vera*. The use of aloe in therapeutic is reported by several scientists (Cera *et al.*, 1980; Aggarwal and Barna 2004; Bairu *et al.*, 2009). A gel in the leaves makes an excellent treatment for wounds, burns and other skin disorders, placing a protective coat over the affected area, speeding up the rate of healing and reducing the risk of infection. Natural propagation of *Aloe vera* is primarily by means of axillary shoots and it is rather a slow way of multiplication to meet the growing demand (Natali *et al.*, 1990a).

2.2.2 *Euphorbia hirta* L.

Euphorbia hirta Linn. is one of such herbs belonging to the family Euphorbiaceae which is frequently seen occupying open waste spaces and grasslands, road sides and pathways. It is usually erect, slender-stemmed, spreading up to 45cm tall, though sometimes can be seen lying down (Singh *et al.*, 2011). The plant is an annual broad-leaved herb that has a hairy stem with many branches from the base to the top. The stem and leaves produce white or milky juice when cut.

This herb is used as an antispasmodic, antiasthmatic, expectorant, anticatarrhal and antisyphilitic (Singh *et al.*, 2005; Singh *et al.*, 2006; Youssouf *et al.*, 2007; Singh *et al.*, 2011). .E. hirta is commonly called asthma weed in Asia and Australia, the herb is widely used in traditional medicine to treat a variety of diseased conditions including asthma, coughs, diarrhea and dysentery (Parekh *et al.*, 2005; Ogbulie *et al.*, 2007).

2.2.3 *Kalanchoe pinnata* (lam.) pers.

Kalanchoe pinnata commonly known as Patharchatta is a genus of Bryophyllum of the family Crassulaceae is a valuable medicinal as well as ornamental plant. It is a genus of about 125 species of tropical region that grows 3-5 feet high. These plants are cultivated as ornamental house plants and rock or succulent garden plants (Kulka 2006). It has a definite ornamental value (Zamora *et al.*, 1998) because of its beautiful inflorescence. It is a popular houseplant and has become naturalized in temperate regions of Asia, the Pacific and Caribbean, Australia, New Zealand, West Indies, Macaronesia, Mascarenes, Galapagos, Melanesia, Polynesia and Hawaii (Zamora *et al.*, 1998).

It is used as ornamental potted plant around the globe. Along with its ornamental values these species also contain a lot of medicinal values (Quisumbing 1951; Joseph *et al.*, 2011). In traditional medicine, these species have been used to treat ailments such as infections, rheumatism and inflammation. Kalanchoe extracts also have immunosuppressive effects.

They are popular because of their ease of propagation, low water requirements, and wide variety of flower colors typically borne in clusters well above the vegetative growth. It is also called as the "Air plant". The "Air plant" *Kalanchoe pinnata* is a curiosity because new individuals develop vegetatively at indents along the leaf, usually after the leaf has broken off the plant and is laying on the ground, where the new plant can take root.

It has a unique ability to sprout baby plants along its leaf edges, which upon contact with soil sprout plants profusely. Virtually any part of the plant will successfully root with great effectiveness.

2.2.4 *Mimosa pudica* L.

Mimosa pudica L. (Mimosaceae) (also called sensitive plant, sleepy plant and the touch-me-not) is a creeping annual or perennial herbaceous species belongs to the family Fabaceae, native to Brazil and largely naturalized throughout the world.

Mimosa pudica is known to contain indoleamines and its alkaloid, mimosine, (3-hydroxy-4-pyridine-1-yl L-alanine) present in leaves and stem (Kokane *et al.*, 2009). The toxic alkaloid mimosine, which has been found to also have antiproliferative and apoptotic effects Aqueous extracts of the roots of the plant have shown significant neutralizing effects in the lethality of the venom of the monocled cobra (Naja Kaouthia) (Pithayanukul *et al.*, 2005).

Green parts of the plant are used as an analgesic, antispasmodic, anti-asthmatic, a mild sedative and anti-depressant. In India it is used as forage for cattle, and is considered to lead to high meat and milk production

In cultivation, this plant is most often grown as an indoor annual, but is also grown for groundcover. It grows mostly in shady areas, under trees or shrubs. Propagation is generally by seed. *Mimosa pudica* can form root nodules that are habitable by nitrogen fixing bacteria (Witkus and Berger 1947). The bacteria are able to convert atmospheric nitrogen, which plants cannot use, into a form that plants can use.

2.2.5 *Rauvolfia serpentina* (L.) BENTH. EX KURZ

Rauvolfia serpentina, or 'Indian snakeroot' or 'sarpagandha' is a species of flowering plant in the family Apocynaceae. It is native to the Indian Subcontinent and East Asia (from India to Indonesia).

Rauvolfia serpentina contains a number of bioactive chemicals, including yohimbine, reserpine, ajmaline, deserpidine, rescinnamine ,serpentinine (Balandrin *et al.*, 1993). The chemical reserpine is an alkaloid first isolated from roots of *R. serpentina* and is used to treat hypertension, blood pressure and mental disorders (Vida 1952)). The roots are bitter, acrid, laxative, thermogenic, diuretic and possess sedative properties. It is highly reputed for hypertension and is useful in stangury, fever, wounds, insomnia, epilepsy and dyspepsia). The decoction of root is used to increase uterine contractions. The juice of the leaf is used as a remedy for the removal of opacities of the cornea. Root is also considered as anthelmintic and antidote to snake venom. Due to indiscriminate and unsustainable harvesting, Rauwolfia has attained the status of "red" listed plant in India (Reddy and Reddy 2008)). IUCN has also kept this plant under endangered status.

In view of the increasing demand, there is a need to develop approaches for efficient propagation. Poor seed viability, low seed germination rate, and enormous genetic variability are the major constraints for the commercial cultivation of *Rauvolfia serpentina* through conventional mode. Whereas vegetative propagation through stem and root cuttings leads to destructive harvesting. In view of this, there is an urgent need to develop *in vitro* methods for the micro propagation and conservation of this valuable endangered medicinal plant.

2.2.6 *Saraca asoca* (ROXB.) DE WILDE

Saraca asoca (the Ashoka tree; lit., "sorrow-less") is a plant belonging to the Caesalpinioideae subfamily of the legume family. Its use in treatment of excessive uterine bleeding is extensive in India. The plant is used also in dysmenorrhea and for depression in women. Ashoka (*Saraca asoca*) has high medicinal value and is used in many Ayurvedic drugs. The International Union for Conservation of Nature and Natural Resources (IUCN) describes the species as "globally vulnerable" (Kingston *et al.*, 2009). The crop can be naturally propagated by seeds and stem grafting. The seedlings are planted in the well manured field during the rainy season.

2.2.7 *Trigonella foenum-graecum* L.

Fenugreek is much used in herbal medicine that belongs to family Fabaceae. The seeds are very nourishing and are given to convalescents and to encourage weight gain, especially in anorexia nervosa. An essential oil is obtained from the seed - used as a food flavouring and medicinally. Fenugreek (*Trigonella foenum-graecum* L.) is an annual legume crop, due to its spice possessing amazing therapeutic and medical properties it is used in many parts of the world. It is one of the oldest medicinal plants known and has long been recognized as a traditional medicine in Asia, (Raju *et al.*, 2001; Bin-Hafeez *et al.*, 2003)).

Fenugreek (*Trigonella foenum-graecum* L.) is an annual forage legume and an important aromatic medicinal plant that is commercially used as spices in major parts of the world. The plant is known for its superior qualities of pharmaceutically and nutraceutically rich in several phytochemicals including several important complex carbohydrates, amino acids, steroids and saponins. The commercial extraction of diosgenin, an important phytochemical from fenugreek is an attractive proposition. However, it is important to make it economically viable by increasing the diosgenin content with genetic, agronomic and biotechnological methods and by reduction in the cost of production for better commercial returns. This could possibly result in an attractive economic return for fenugreek growers and producers. In addition to common breeding approaches, modern tissue culture or *in vitro* techniques offers an important opportunity to improve the plant properties via genetic engineering and has recently been used as a tool for genetic transformations.

Table 1: Scientific classification of medicinal plants selected in present research.(Source: http://en.wikipedia.org/)

Aloe vera		Euphorbia hirta		Kalanchoe pinnata		Mimosa pudica		Rauvolfia serpentina		Ashoka tree		Fenugreek	
Scientific classification													
Kingdom:	Plantae	Kingdom:	Plantae	Kingdom:	Plantae	Kingdom:	Plantae	Kingdom	Plantae	Kingdom:	Plantae	Kingdom:	Plantae
Clade:	Angiosperms	(unranked):	Angiosperms	(unranked):	Angiosperms	(unranked):	Angiosperms	Division:	Magnoliophyta	(unranked):	Angiosperms	(unranked):	Angiosperms
Clade:	Monocots	(unranked):	Eudicots	(unranked):	Eudicots	(unranked):	Eudicots	Class:	Magnoliopsida	(unranked):	Eudicots	(unranked):	Eudicots
Order:	Asparagales	(unranked):	Rosids	(unranked):	Core eudicots	(unranked):	Rosids	Order:	Gentianales	(unranked):	Rosids	(unranked):	Rosids
Family:	Xanthorrhoeaceae	Order:	Malpighiales	Order:	Saxifragales	Order:	Fabales	Family:	Apocynaceae	Order:	Fabales	Order:	Fabales
Subfamily:	Asphodeloideae	Family:	Euphorbiaceae	Family:	Crassulaceae	Family:	Fabaceae	Genus:	Rauvolfia	Family:	Fabaceae	Family:	Fabaceae
Genus:	Aloe	Genus:	Euphorbia	Genus:	Bryophyllum	Subfamily:	Mimosoideae	Species:	R. serpentina	Genus:	Saraca	Genus:	Trigonella
Species:	A. vera	Species:	E. hirta	Species:	B. pinnata	Genus:	Mimosa			Species:	S. asoca	Species:	T. foenum-graecum
						Species:	M. pudica						

3 REVIEW OF LITERATURE

India is sitting on a gold mine of well-recorded and traditionally well-practised knowledge of herbal medicine, called the botanical garden of the world. Medicinal plants and herbs are used by nearly all cultures to prevent or treat illness. Many countries like Africa, Asian countries nearly 80% of the population depends on herbal medicines as a primary source of care. Medicinal plants consider as a rich resources of ingredients, which can be used in drug and synthesis. In recent years, there has been renewed interest in natural medicines that are obtained from plant parts or plant extracts. Herbal garden is being used as resource point for identification and advocating the uses of medicinal plant.

Consumption of herbal medicines is widespread and increasing. Harvesting from the wild, the main source of raw material, is causing loss of genetic diversity and habitat destruction (Watson 1947). Domestic cultivation is a viable alternative and offers the opportunity to overcome the problems that are inherent in herbal extracts: misidentification, genetic and phenotypic variability, extract variability and instability, toxic components and contaminants (Farnsworth and Soejarto 1991; Briskin 2000). The use of controlled environments can overcome cultivation difficulties and could be a means to manipulate phenotypic variation in bioactive compounds and toxins. Conventional plant-breeding methods can improve both agronomic and medicinal traits, and molecular marker assisted selection will be used increasingly. There has been significant progress in the use of tissue culture and genetic transformation to alter pathways for the biosynthesis of target metabolites. Obstacles to bringing medicinal plants into successful commercial cultivation include the difficulty of predicting which extracts will remain marketable and the likely market preference for what is seen as naturally sourced extracts (Canter *et al.*, 2005).

As species face loss of habitat or overharvesting, there have been new issues to deal with in sourcing crude drugs (Foster 1992; Brooks *et al.*, 2002). These include changes to the herb from farming practices, substitution of species or other plants altogether, adulteration and cross-pollination issues(Canter *et al.*, 2005) .

Well-developed techniques are currently available to help growers meet the demand of the pharmaceutical industry in the next century. These protocols are designed to provide optimal levels of carbohydrates, organic compounds (vitamins), mineral nutrients, environmental factors (e.g. light, gaseous environment, temperature, and humidity) and growth regulators required to obtain high regeneration rates of many plant species *in vitro* and thereby facilitate commercially viable micropropagation (Aitken-Christie *et al.,* 1995; DiCosmo and Misawa 1995). Well-defined cell culture methods have also been developed for the production of several important secondary products (Rout *et al.,* 2000).

Experimental approaches used for propagation of medicinal plants through tissue culture can be divided into three broad categories. The most common approach is to isolate organised meristems like shoot tips or axillary buds and induce them to grow into complete plants. This system of propagation is commonly referred to as micropropagation. In the second approach, adventitious shoots are initiated on leaf, root and stem segments or on callus derived from those organs. The third system of propagation involves induction of somatic embryogenesis is in cell and callus cultures. This system is theoretically most efficient as large numbers of somatic embryos can be obtained once the whole process is standardised.

Biotechnology involving modern tissue culture, cell biology and molecular biology offers the opportunity to develop new germplasm that are well adapted to changing demands.

Medicinal plant based industries and businesses rely mostly on the source of raw material coming from fields where these plants are being propagated by natural means mostly through seeds. As seed based propagation always carries variations thus, overall yield from such a source results a great loss to the businesses.

Many Indian plants are being examined for their beneficial uses and reports occur in numerous scientific journals of the merits of using such plants (Pitchai *et al.,* 2010; Daisy *et al.,* 2011; Tota *et al.,* 2013). Many previous efforts were made to compile the information of the reported medicinal plants and develop a database bud were not comprehensive. Here we are proposing to develop a database of Indian medicinal plants to provide a single source of comprehensive literature about the features, habitat, propagation and uses.

Medicinal plants selected for present research work were individually reviewed for their natural and *in vitro* propagation related research work and compiled in the following section.

3.1.1 *Aloe barbadensis* **MILL.**

Aloe barbadensis has been cultured through various explants by many researcher *viz.* shoot tip (Oliveira *et al.,* 2008), seed (Bairu *et al.,* 2009) and by leaf pulp (Grindlay and Reynolds 1986). We use mainly regional mature leaf for tissue culture of *A. vera.* The present research finding shows the promising results for regeneration of aloe from mature leaf explants. Further, the investigation was carried out to establish the callus culture protocol for mass multiplication of callus from leaf explants. This can be used for large-scale production of active metabolite anthraquinone (used as a laxative inflammation) of aloe directly from callus cultures in industries.

3.1.2 *Euphorbia hirta* **L.**

E. hirta has been used to treat asthma as a folk medicine (Higashi *et al.*, 1991). *E. hirta* functions for the treatment of asthma is probably through synergistic anti-inflammatory and antioxidant activities of especially the flavonoids, sterols and triterpenoids (Ponou *et al.*, 2008). Asthma has long been associated with chronic inflammation and an overall increase in reactive groups and oxidative stress (Shih and Cherng 2012). *E. hirta* also existed significant activity to prevent early and late phase allergic reactions and thereby asthma. No significant reports are available on micropropagation of *E. hirta*, this led us to work on *in vitro* regeneration of this medicinal plant through callus regeneration and organogenesis which also opens up new avenues for commercial exploitation of medicinal properties of this highly valuable medicinal plant.

3.1.3 *Kalanchoe pinnata* **(lam.) pers.**

As it produce valuable secondary metabolites having different types of medicinal importance, so it is necessary to develop protocols for rapid multiplication through micropropagation, callogenesis and organogenesis, because efficient regeneration system is prerequisite for all biotechnological tools. (Gabryszewska and Hempel 1984) reported regeneration of roots in Kalanchoe first time A few other species of Bryophyllum have been regenerated e.g., *K. daigremontianum* and *K. laciniata* (Shagufta Naz *et al.*, 2009) for the purpose of finding an effective protocol. Multiplication and growth response were obtained on a hormone free MS basal medium suggesting that there is a little role of plant growth hormones in the *in vitro* development, multiplication and organogenesis of Kalanchoe (Khan *et al.*, 2006). Thus, *in vitro* propagation could be a valuable alternative to propagation by seeds or cuttings, especially when breeders want to work with species hybrids.

3.1.4 *Mimosa pudica* **L.**

This plant is useful in treatment of various diseases. Therefore, it is important to perform micropropagation of this plant. Recent studies shown that when cotyledons, hypocotyls, leaflets and shoot apices of *Mimosa pudica*, cultured on B_5 medium comprising of an auxin, 2,4-D (0.5 ppm), and a cytokinin, BAP (1 ppm), showed high percentage of callusing in all the explants within a few weeks of culture. After about 2 months of culture, multiple shoot buds differentiated from these which could be successfully rooted on BM or BM + IAA (2 ppm) to form complete plantlets (Gharyal and Maheshwari 1982)

3.1.5 *Rauvolfia serpentina* (L.) BENTH. EX KURZ

In vitro shoots multiplied on Murashige and Skoog basal liquid nutrients supplemented with BAP (1.0 mg/L) and NAA (0.1 mg/L) and in-vitro rhizogenesis was observed in modified MS basal nutrient containing NAA (1.0 mg/L) and 2% sucrose by Goel *et al.,* (2009). Multiple shoots induced from nodal segments and shoot apices of *Rauvolfia serpentina* on MS medium containing 1.0 mg/l BA and 0.1 mg/l NAA was obtained by Sarker *et al.,* (1996).

3.1.6 *Saraca asoca* (ROXB.) DE WILDE

Ashoka (*Saraca asoca*), it is grown as ornamental tree throughout the country; in Western Ghats of Tamil Nadu. The bark is extremely useful in gyaecological problems especially, uterine bleeding associated with fibroids. *In vitro* clonal propagation of Ashoka, *Saraca asoca* (Roxb) De Wilde plant was achieved using shoot tip, nodal and intermodal explants (Shahid *et al.,* 2007; Waman *et al.,* 2008). MS medium supplemented with different concentrations (0.5-2.0 mg/l) of Benzyl amino purine (BAP), Kinetin (Kn) and 2, 4-Dichlorophenoxy acetic acid (2,4-D). The frequency of shoot organogenesis was highest (82%) in nodal explants treated with 0.5 mg/l of BAP and callus were formed more on 2, 4-D. The micro-shoots rooted well on MS medium supplemented with 4.0 mg/l of IBA. Hardened regenerants (40%) were acclimatized to the soil (Subbu *et al.,* 2008).

3.1.7 *Trigonella foenum-graecum* L.

Fresh and dried leaves and seeds used by Oncina *et al.,* (2000), for production of secondary metabolites. Fenugreek seed is an important source of steroidal sapogenins such as diosgenin, which are used extensively by both pharmaceutical, and nutraceutical industries.

Callus generation increase was caused by the increase concentration of 1 mg/l 2,4-D hormone used by Rezaeian (2011).

4 MATERIAL AND METHODS

4.1 ESTABLISHMENT OF *IN VITRO* CULTURES OF ELITE MEDICINAL PLANTS

4.1.1 Plant Material

The planting material of medicinal plants were collected from different nurseries, seed depositories and botanical gardens etc. For the establishment of *in vitro* cultures of different medicinal plants the explants viz. nodal segments, leaves, root etc. were collected from the plants of superior genotypes.

4.1.2 Surface sterilization

The various explants were subjected to surface sterilization by treatment with Bavistin 1% solution for 45 minutes followed thoroughly rinsing with sterilized water. The other sterilizing agents (mercuric chloride, hypo solution and alcohol) were also used after standardizing the protocol of sterilization for different explants of different plant species.

4.1.3 Establishment of *in vitro* cultures

For the establishment of *in vitro* cultures of medicinal plants, the different pathways viz. Axillary bud induction, callogenesis and organogenesis, were tested.

4.1.4 Standardizing physico-chemical conditions

The different micro propagation pathways were formulated by standardizing the media type along with growth regulator.

MS basal medium was used in all experiments supplemented with different permutations of cytokinins (BAP and Kinetin) auxins (2,4-D and NAA).

4.1.5 Statistical Analysis

Each treatment was performed in replicates. The mean and standard deviations were computed from each treatment.

4.2 DEVELOPING THE MEDICINAL PLANT DATABASE

4.2.1.1 *Data collection*

Data of medicinal plants for which the tissue culture has been practiced was collected from literature sources such as PubMed, Science Direct, Biomed Central, Wiley journals, etc. For the collection of data, an online form was generated followed by storing the information in MS Access 2013.

4.2.1.2 Database Design

The online database was constructed using google site. The html pages of each medicinal plant was created and designed individually.

5 RESULTS AND DISCUSSION

5.1 ESTABLISHMENT OF *IN VITRO* CULTURES OF ELITE MEDICINAL PLANTS

5.1.1 *Aloe barbadensis* MILL.

In vitro propagation of aloe has been reported earlier through Shoot tip, axillary bud, meristem, underground stem and inflorescence have been used as explant (Natali *et al.*, 1990b; Abrie and Van Staden 2001; Aggarwal and Barna 2004). Using axillary buds as the explants has been proven to be the most successful and efficient micropropagation procedure for aloe yet (Roy and Sarkar 1991).

In present study, we used mature leaf for tissue culture of *A. vera*. Callus induction was obtained from mature leaf explants when cultured on MS medium supplemented with (1-2 mg/l) BAP and (1-2 mg/l) 2,4-D (Graph 1, Figure 1). Contrary to our research Monge *et al.*, 2008 reported the initiation of embryogenic callus form three leaf base explants on MS + 2.5 mg/l 2,4-D, 2 mg/l BAP, and 40 mg/l adenine sulphate.

Moderate callusing was obtained when leaf discs were subcultured on MS (basal medium) supplemented with (1-2) mg/l BAP, and (1-2) mg/l 2,4-D. Efforts were also made to regenerate callus on activated charcoal containing medium (Figure 2). (Data not given)

The present research finding shows the promising results for regeneration of aloe from mature leaf explants. Further, the investigation was carried out to establish the callus culture protocol for mass multiplication of callus from leaf explants. This can be used for large-scale production of active metabolite anthraquinone (used as a laxative inflammation) of aloe directly from callus cultures in industries.

5.1.2 *Euphorbia hirta* L.

The responding leaf explants showed visible signs of swelling within one week of culture. Callus induction from leaf and seed explants was readily obtained within two weeks of incubation when auxins in combination with cytokinins of different concentrations were used. Heavy callogenic response from leaves was obtained when cultured on MS media supplemented with 1 mg/l NAA and 1 mg/l 2,4-D (Figure 3 & 4 and Graph 2).

5.1.3 *Kalanchoe pinnata* (lam.) pers.

The presented study was targeted to establish an efficient, fast and reliable protocol for the callogenesis and shoot regeneration of the *Kalanchoe pinnata* plant from the leaf explants.

The disinfection treatment used was efficient for the *in vitro* establishment of *Kalanchoe pinnata*, with approximately 90% of explants remaining aseptic and showing new growth after 20 days of culture (Figure 5). To induce shoot regeneration, individual leaves of at least 1x1cm^2 of plant were transferred to MS medium supplemented with different concentrations of BAP (0.5 – 5.0 mg/l) and maximum number of shoots were obtained on 1mg/l BAP while maximum length of shoots (approx 3cm) was obtained on 2 mg/l BAP. New shoots first appeared as small, green protuberances at the leaf bases and developed into shoots (Graph 4).

To induce callogenesis, individual leaves were transferred to MS medium supplemented with different concentrations of 2, 4,D and NAA and the heavy callus was obtained on 1 mg/NAA and 1 mg/l 2,4-D. Shooting along with rooting was also obtained on hormone free media (Figure 6, Graph 5).

5.1.4 *Mimosa pudica* L.

Different combinations of Auxins (NAA and 2,4-D) were standardized for initiation of shoot and callus from stem explants. Shoot initiation response was obtained on MS media supplemented with (0.5-1) mg/l 2,4-D and 1mg/l NAA. Shoots were successfully initiated established from nodal part (stem) explants and length of shoots were maximum with 0.5 mg/l 2,4-D and 1mg/l NAA after 1 week of inoculation (Figure 7, Graph 6). Higher percentage of callus was obtained with 0.5 mg/l 2,4-D and 1mg/l NAA after 2 weeks of inoculation (Figure 8).

5.1.5 *Rauvolfia serpentina* (L.) BENTH. EX KURZ

In earlier studies regeneration through callusing in leaf and direct regeneration using apical and axillary nodes has been reported by many scientists (Sarker *et al.*, 1996; Goel *et al.*, 2009). Limited efforts were made to regenerate plants from mature plant part through tissue culture. Thus, the present study was undertaken to develop more efficient protocol for *in vitro* propagation system using nodal explant and mature leaf as explants.

Callogenic response was obtained from leaf explants when cultured on MS medium supplemented with (0.2-1 mg/l) NAA + (0.5 mg/l) 2,4-D/ (0.5 mg/l) BAP (Figure 10).

The explants cultured on MS medium supplemented with different concentrations and combination of NAA and BAP showed varied response for shoot formation. The percentage of shoot formation and length of shoot was minimum at lower concentrations of NAA and BAP, the maximum percentage of shoot formation and maximum number of shoots was obtained with 1mg/l NAA and 1mg/l BAP (Figure 9, Graph 7).

5.1.6 *Saraca asoca* (ROXB.) DE WILDE

Callus cultures were successfully obtained established from leaf explant. callogenic response was obtained from leaf explants when cultured on MS medium supplemented with (1-7 mg/l) BAP + (1-4 mg/l) 2,4-D. Curling and swelling of leaf was initiated followed by high callusing on MS media with hormonal concentrations of (5 mg/l) 2,4-D and (5 mg/l) BAP (Figure 11, Figure 12 and Graph 8).

5.1.7 *Trigonella foenum-graecum* L.

The medicinal plant that is used in our report is highly rich in steroidal sapogenin, diosgenin and its major alkaloid, trigonelline. Callogenesis was initiated from the young leaf explant of fenugreek. Callus generation was caused by the increase concentration of 1mg/l 2,4-D hormone when used by Rezaeian (2011).

In our study, the three explants (flowers, leaves , stem) of this plant were cultured on MS medium supplemented with BAP (1-5 mg/l) + 2,4-D (1-5 mg/l) . The callus was successfully initiated on flowers and leaves explants in combined concentrations of both hormones having same amount only and were poorly initiated when stem was taken as explant on same media and hormone (Figure 13, Figure 14, Graph 9 and Graph 10).

5.2 DEVELOPING THE MEDICINAL PLANT DATABASE

An online web based static database on Indian medicinal plants was developed using the google site free service (available at https://sites.google.com/site/meddbase).

The current database features different pages dedicated to plant science knowledge, research and allied information. Database was compiled in A to Z in 26 pages containing a list of medicinal plants. Each medicinal plant in the list was linked to another page featuring information about the medicinal properties of plants alongwith the tissue culture / propagation references.

Tissue culture / propagation references were collected from google scholar and may other online line reference databases. The references were collected and compiled in XML format and further linked to site in the form of table.

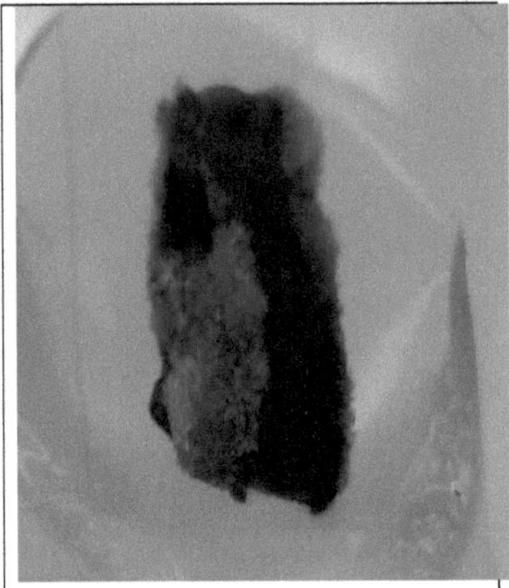

Figure 1: Initiation of Callus from mature leaf explant of Aloe barbadensis

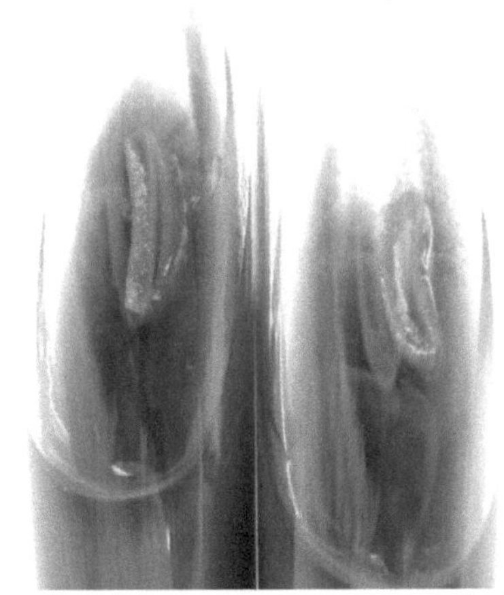

Figure 2: Culture of leaf explant on activated charcoal containing MS medium

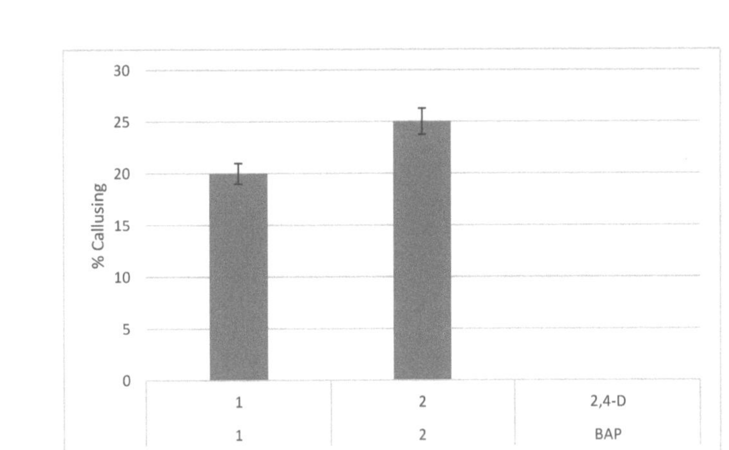

Graph 1: Initiation of callus from leaf explant of Aloe barbadensis, cultured on MS medium. (Data was recorded after 3 weeks)

Figure 3: Initiation of Callus from Seed explant of Euphorbia hirta

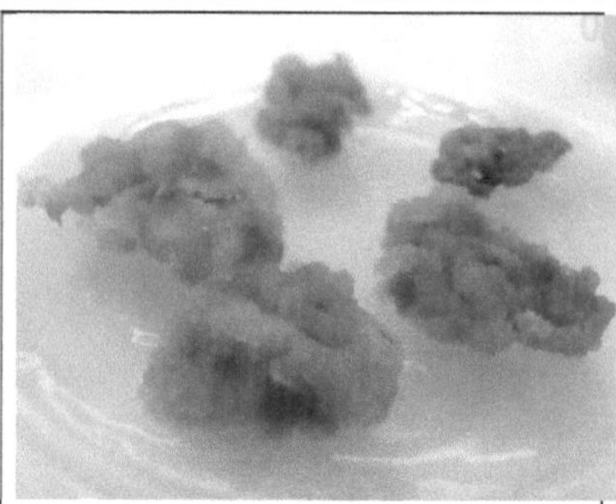

Figure 4: Initiation of Callus culture from leaf explant of Euphorbia hirta.

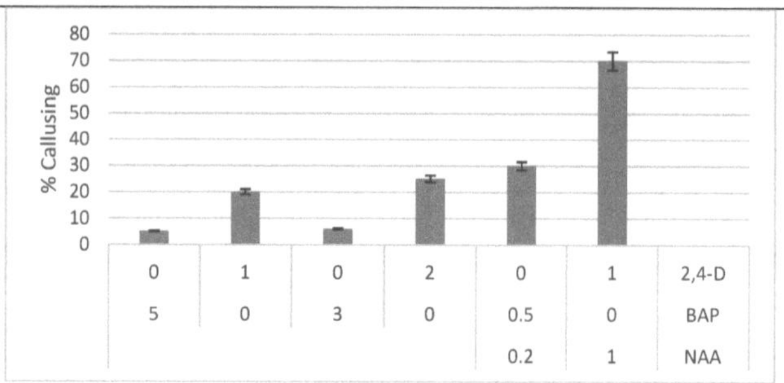

Graph 2:Initiation of callus culture from leaf explant of Euphorbia hirta, cultured on MS medium with different permutations of PGRs.

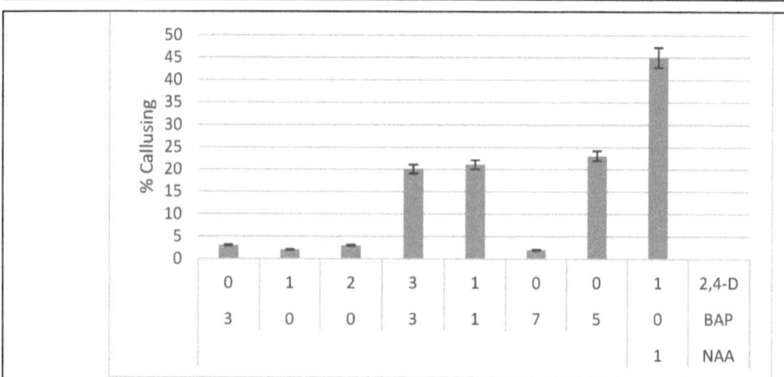

Graph 3: Initiation of callus culture from seed explant of Euphorbia hirta, cultured on MS medium with different permutations of PGRs.

Figure 5: Initiation of shoots from mature leaf explant of Kalanchoe pinnata.

Figure 6: Initiation of callus from leaf explant of Kalanchoe pinnata

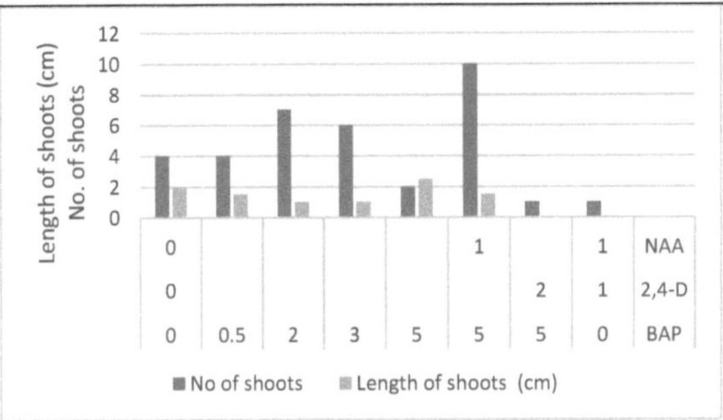

Graph 4: Initiation of shoots from leaf explant of Kalanchoe pinnata, cultured on MS medium with different combinations of BAP, 2,4-D and NAA. (Data was recorded after 4 weeks of culturing)

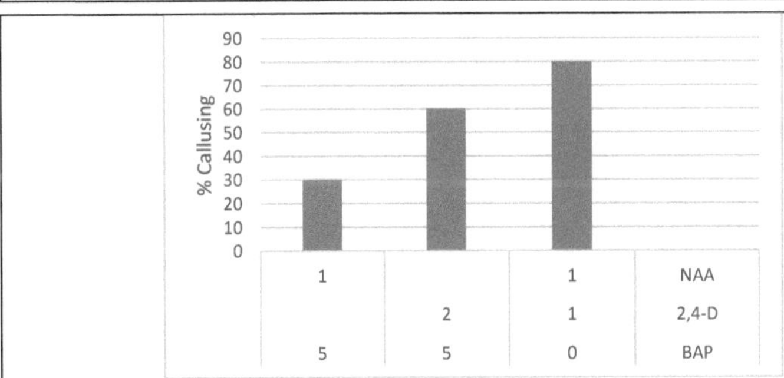

Graph 5: Initiation of callus from leaf explant of Kalanchoe pinnata, cultured on MS medium with different combinations of BAP, 2,4-D and NAA. (Data was recorded after 4 weeks of culturing)

Figure 7: Initiation of shoots from mature nodal explants of Mimosa pudica

Figure 8: Initiation of callus from shoot explant of Mimosa pudica

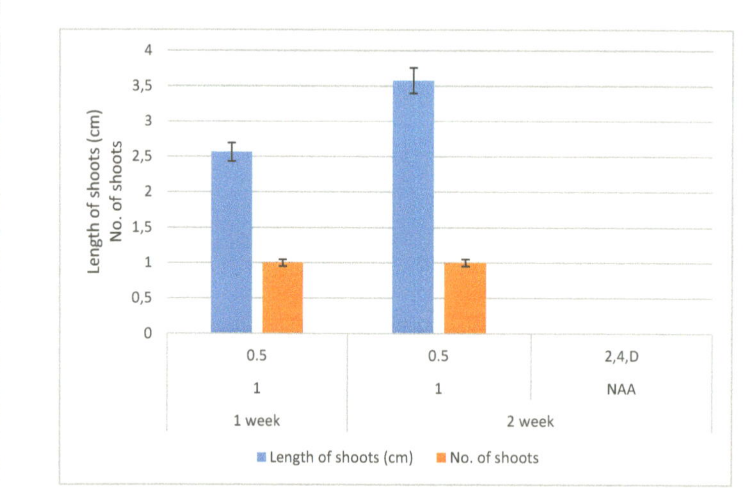

Graph 6: Initiation of shoots from nodal explant of Mimosa pudica, cultured on MS medium with different combinations of NAA and 2,4-D. (Data was recorded after 8 weeks of culturing)

Figure 9: Initiation of buds from nodal explant of Rauvolfia serpentina

Figure 10: Initiation of callus culture from shoot discs as explant of Rauvolfia serpentina

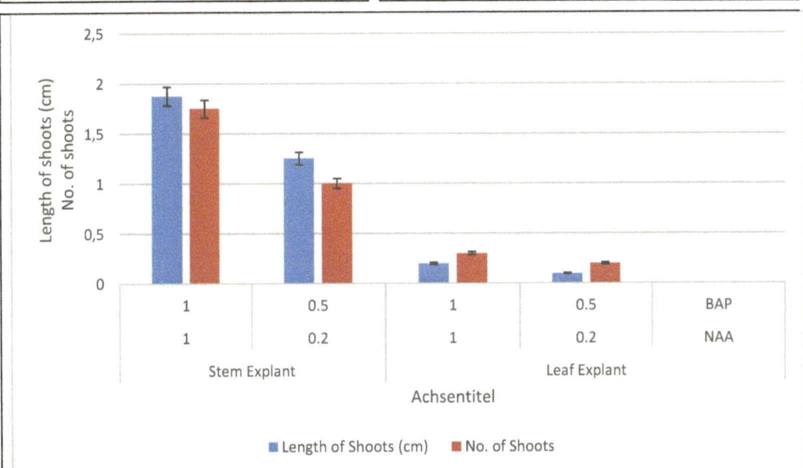

Graph 7: Initiation of shoots on nodal explants of Rauvolfia serpentina, cultured on MS medium with different combinations of BAP and NAA. (Data was recorded after 4 weeks)

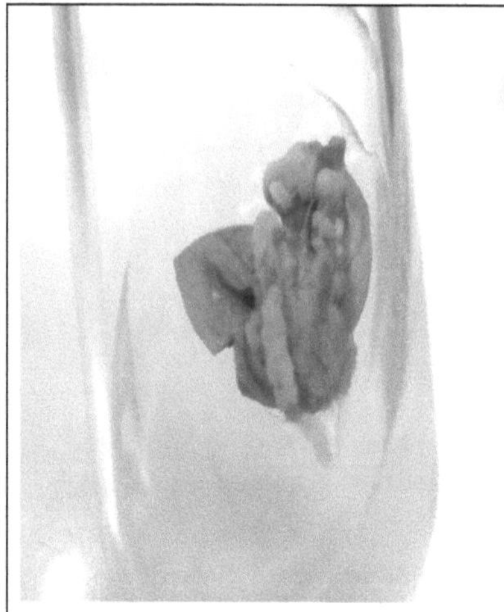

Figure 11: Initiation of Callus from young leaf explant of Saraca asoca

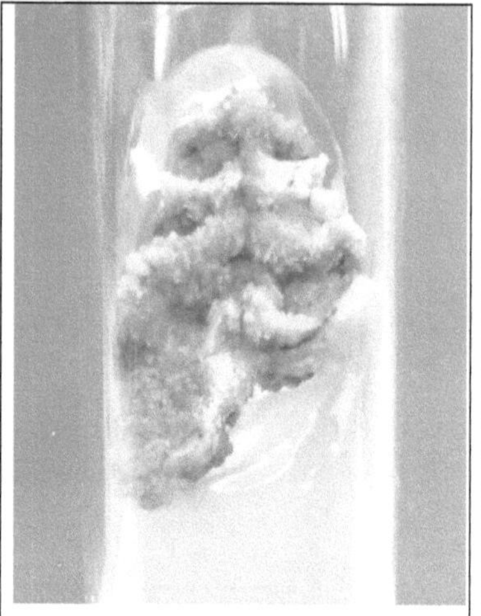

Figure 12: Initiation of callus from mature leaf explant of Saraca asoca

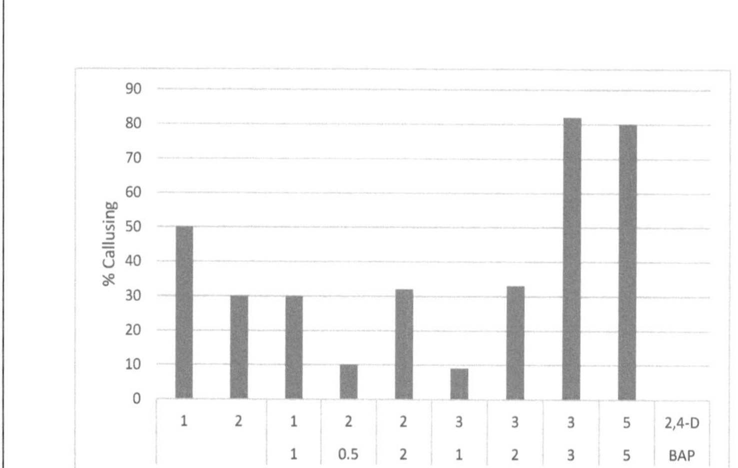

Graph 8: Initiation of callus from leaf explant of Saraca asoca, cultured on MS medium supplemented with different combinations of BAP and 2,4-D. (Data was recorded after 4 weeks)

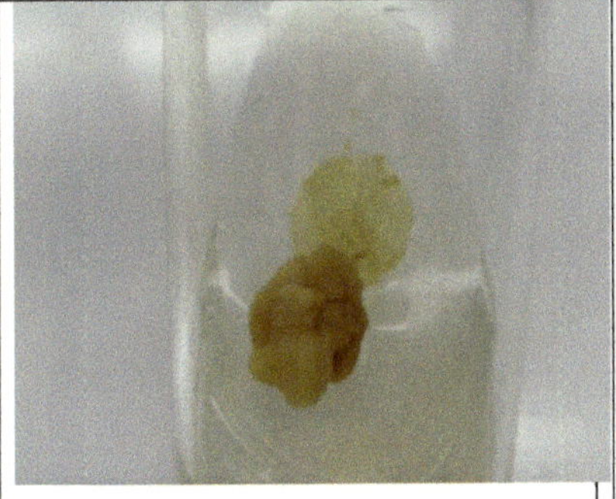

Figure 13: Initiation of callus from leaf explant of Trigonella foenum.

Figure 14: Initiation of callus from flower of Trigonella foenum.

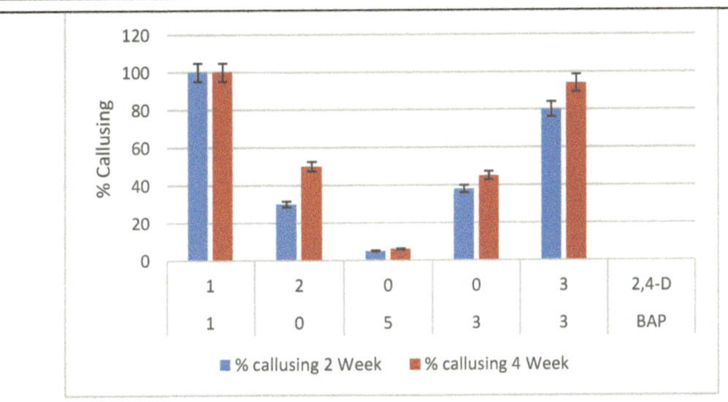

Graph 9: Initiation of callus from flower of methi cultured on MS medium supplemented with BAP and 2,4-D. (Data was recorded after 2ⁿᵈ week and 4ᵗʰ week).

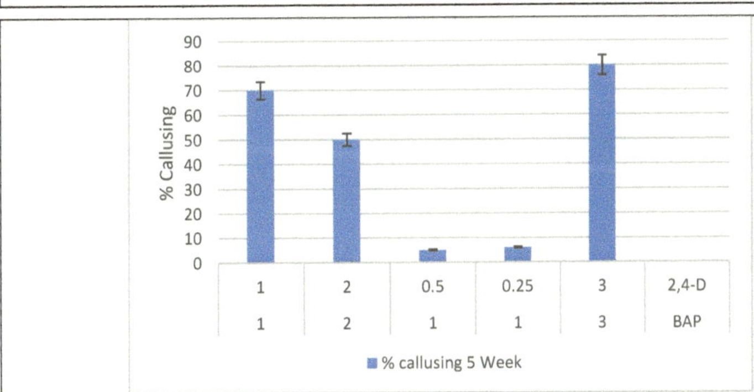

Graph 10: Initiation of callus from mature leaves of methi cultured on MS medium supplemented with BAP and 2,4-D. (Data was recorded after 5 weeks)

Figure 15: Home page of MedDBase (Medicinal Plant Database), featuring different pages as Home, Discover plants, database, about us, blog and contact us.

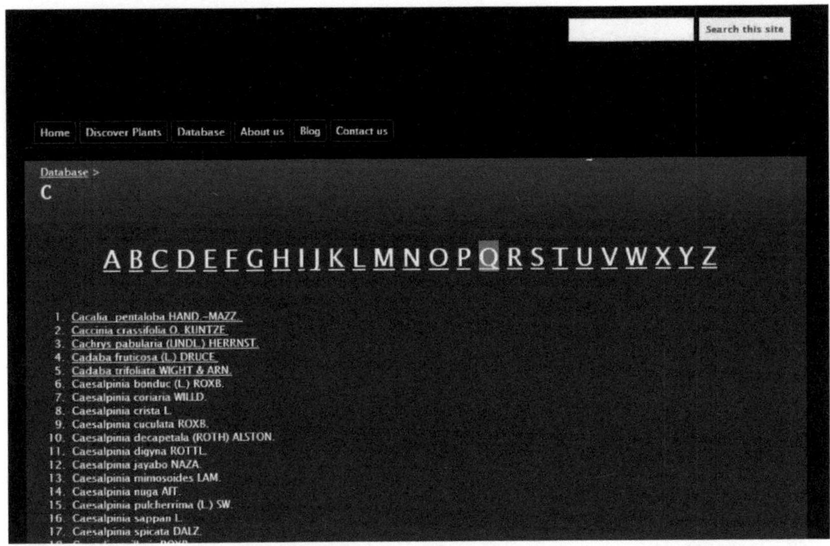

Figure 16: Database arranged from A to Z alphabetical order, each letter containing a list of medicinal plants.

Figure 17: The information page of medicinal plant.

QuickSearch: [ ] clear Number of matching entries: 0.

Author	Title	Year	Journal/Proceedings	Reftype	DOI/URL
Adaniya, S. & Shirai, D.	In vitro induction of tetraploid ginger (textless itextgreater Zingiber officinaletextless/itextgreater Roscoe) and its pollen fertility and germinability	2001	Scientia horticulturae Vol. 88(4), pp. 277â€"287	article	URL
Arimura, C.T., Finger, F.L. & Casali, V.W.D.	Effect of ANA and BAP on ginger (Zingiber officinale Roscoe) sprouting in solid and liquid medium.	2000	Revista Brasileira de Plantas Medicinais Vol. 2(2), pp. 23â€"26	article	URL
Babu, K.N., Lukose, R., Ravindran, P.N. & Vidhyasekaran, P.	Tissue culture of tropical spices.	1993	Genetic engineering, molecular biology and tissue culture for crop pest and disease management., pp. 257â€"267	article	URL
Bajaj, Y.P.S., Furmanowa, M. & Olszowska, O.	Biotechnology of the micropropagation of medicinal and aromatic plants	1988	Medicinal and aromatic plants I, pp. 60â€"103	incollection	URL
Dekkers, A.J., Rao, A.N. & Goh, C.J.	In vitro storage of multiple shoot cultures of gingers at ambient temperatures of 24â€"29 C	1991	Scientia horticulturae Vol. 47(1), pp. 157â€"167	article	URL
Faria, R.T. & Illg, R.D.	Micropropagation oftextless itextgreater Zingiber spectabiletextless/itextgreater Griff.	1995	Scientia horticulturae Vol. 62(1), pp. 135â€"137	article	URL
Guan, Q.Z., Guo, Y.H., Sui, X.L., Li, W. & Zhang, Z.X.	Changes in photosynthetic capacity and antioxidant enzymatic systems in micropropagated Zingiber officinale plantlets during their acclimation	2008	Photosynthetica Vol. 46(2), pp. 193â€"201	article	URL
Guo, Y. & Zhang, Z.	Establishment and plant regeneration of somatic embryogenic cell suspension cultures of thetextless itextgreater Zingiber officinaletextless/itextgreater Rosc.	2005	Scientia horticulturae Vol. 107(1), pp. 90â€"96	article	URL
HIREMATH, RAJANI.C.	Micropropagation of ginger (Zingiber officinale Rosc.)	2006	*School:* UNIVERSITY OF AGRICULTURAL SCIENCES	phdthesis	URL
Jasrai, Y.T., Patel, K.G. & George, M.M.	Micropropagation of Zingiber officinale Rosc. and Curcuma amada Roxb.	2000	Spices and aromatic plants: challenges and opportunities in the new century. Contributory papers. Centennial conference on spices and aromatic plants, Calicut, Kerala, India, 20–23 September, 2000., pp. 52â€"54	inproceedings	URL
Kun-Hua, W., Jian-Hua, M., He-Ping, H. & Shan-Lin, G.	Generation of autotetraploid plant of ginger (Zingiber officinale Rosc.) and its quality evaluation.	2011	Pharmacognosy magazine Vol. 7(27), pp. 200â€"206	article	URL
Lincy, A.K., Jayarajan	Relationship between vegetative and	2008	Scientia horticulturae	article	URL

Figure 18: References related to propagation and tissue culture work done in past on particular medicinal plant species.

6 REFERENCES

Abrie AL, Van Staden J. 2001. Micropropagation of the endangered *Aloe polyphylla*. *Plant Growth Regulation* **33**: 19–23.

Aggarwal D, Barna KS. 2004. Tissue culture propagation of elite plant of *Aloe vera* Linn. *Journal of Plant Biochemistry and Biotechnology* **13**: 77–79.

Aitken-Christie J, Kozai T, Takayama S. 1995. Automation in plant tissue culture. General introduction and overview. *Automation and environmental control in plant tissue culture*: 1–18.

Bairu MW, Kulkarni MG, Street RA, Mulaudzi RB, Van Staden Johannes. 2009. Studies on seed germination, seedling growth, and *in vitro* shoot induction of *Aloe ferox* Mill., a commercially important species. *HortScience* **44**: 751–756.

Balandrin MF, Kinghorn AD, Farnsworth NR. 1993. Plant-derived natural products in drug discovery and development. *Human Medicinal Agents from Plants. American Chemical Society, Washington, DC*.

Billing J, Sherman PW. 1998. Antimicrobial functions of spices: why some like it hot. *Quarterly Review of Biology*: 3–49.

Bin-Hafeez B, Haque R, Parvez S, Pandey S, Sayeed I, Raisuddin S. 2003. Immunomodulatory effects of fenugreek (*Trigonella foenum* graecum L.) extract in mice. *International immunopharmacology* **3**: 257–265.

Briskin DP. 2000. Medicinal plants and phytomedicines. Linking plant biochemistry and physiology to human health. *Plant Physiology* **124**: 507–514.

Brooks TM, Mittermeier RA, Mittermeier CG, Da Fonseca GA, Rylands AB, Konstant WR, Flick P, Pilgrim J, Oldfield S, Magin G. 2002. Habitat loss and extinction in the hotspots of biodiversity. *Conservation biology* **16**: 909–923.

Canter PH, Thomas H, Ernst E. 2005. Bringing medicinal plants into cultivation: opportunities and challenges for biotechnology. *Trends in Biotechnology* **23**: 180–185.

Cera LM, Heggers JP, Robson MC, Hagstrom WJ. 1980. The therapeutic efficacy of *Aloe vera* cream (Dermaide Aloe) in thermal injuries: two case reports [Dogs]. *Journal American Animal Hospital Association*.

Daisy P, Singh Sanjeev Kumar, Vijayalakshmi P, Selvaraj C, Rajalakshmi M, Suveena S. 2011. A database for the predicted pharmacophoric features of medicinal compounds. *Bioinformation* **6**: 167–168.

DiCosmo F, Misawa M. 1995. Plant cell and tissue culture: alternatives for metabolite production. *Biotechnology advances* **13**: 425–453.

Fabricant DS, Farnsworth NR. 2001. The value of plants used in traditional medicine for drug discovery. *Environmental health perspectives* **109**: 69.

Farnsworth NR, Soejarto DD. 1991. Global importance of medicinal plants. *The conservation of medicinal plants*: 25–51.

Foster S. 1992. Medicinal plant conservation and genetic resources: examples from the temperate northern hemisphere. *WOCMAP I-Medicinal and Aromatic Plants Conference: part 4 of 4 330.67–74.*

Gabryszewska E, Hempel M. 1984. The influence of cytokinins and auxins on Alstroemeria in tissue culture. *II Symposium on Growth Regulators in Floriculture 167.295–300.*

Gharyal PK, Maheshwari SC. 1982. Plantlet Formation in Tissue Cultures of the Sensitive Plant *Mimosa pudica* L. *Zeitschrift für Pflanzenphysiologie* **105**: 179–182.

Goel MK, Mehrotra S, Kukreja AK, Shanker K, Khanuja SPS. 2009. *In vitro* propagation of Rauwolfia serpentina using liquid medium, assessment of genetic fidelity of micropropagated plants, and simultaneous quantitation of reserpine, ajmaline, and ajmalicine. *Protocols for In vitro Cultures and Secondary Metabolite Analysis of Aromatic and Medicinal Plants*. Springer, 17–33.

Grindlay D, Reynolds T. 1986. The *Aloe vera* phenomenon: A review of the properties and modern uses of the leaf parenchyma gel. *Journal of ethnopharmacology* **16**: 117–151.

Higashi O, Takagaki T, Koshimizu K, Watanabe K, Kaji M, Hoshino J, Nishida T, Huffman MA, Takasaki H, Jato J. 1991. Biological activities of plant extracts from tropical Africa. *African Study Monographs* **12**: 201–210.

Joseph B, Sridhar B, Sankarganesh J, Edwin BT. 2011. Rare medicinal plant-*Kalanchoe pinnata*. *Res J Microbiol* **6**: 322–327.

Khan S, Naz Sheeba, Ali K, Zaidi S. 2006. Direct organogenesis of *Kalanchoe tomentosa* (Crassulaceae) from shoot-tips. *Pakistan Journal of Botany* **38**: 977.

Kingston C, Jeeva S, Jeeva GM, Kiruba S, Mishra BP, Kannan D. 2009. Indigenous knowledge of using medicinal plants in treating skin diseases in Kanyakumari district, Southern India. *Indian journal of traditional knowledge* **8**: 196–200.

Kokane DD, More RY, Kale MB, Nehete MN, Mehendale PC, Gadgoli CH. 2009. Evaluation of wound healing activity of root of *Mimosa pudica*. *Journal of ethnopharmacology* **124**: 311–315.

Kulka RG. 2006. Cytokinins inhibit epiphyllous plantlet development on leaves of *Bryophyllum* (*Kalanchoe*) *marnierianum*. *Journal of experimental botany* **57**: 4089–4098.

Li JW-H, Vederas JC. 2009. Drug discovery and natural products: end of an era or an endless frontier? *Science* **325**: 161–165.

Lichterman BL. 2004. Book: Aspirin: The Story of a Wonder Drug. *BMJ: British Medical Journal* **329**: 1408.

Natali L, Sanchez IC, Cavallini A. 1990a. *In vitro* culture of *Aloe barbadensis* Mill.: Micropropagation from vegetative meristems. *Plant cell, tissue and organ culture* **20**: 71–74.

Natali L, Sanchez IC, Cavallini A. 1990b. *In vitro* culture of *Aloe barbadensis* Mill.: Micropropagation from vegetative meristems. *Plant cell, tissue and organ culture* **20**: 71–74.

Naz Shagufta, Javad S, Ilyas S, Ali A. 2009. An efficient protocol for rapid multiplication of *Bryophyllum pinnata* and *Bryophyllum daigremontianum*. *Pak. J. Bot* **41**: 2347–2355.

Ogbulie JN, Ogueke CC, Okoli IC, Anyanwu BN. 2007. Antibacterial activities and toxicological potentials of crude ethanolic extracts of *Euphorbia hirta*. *African Journal of Biotechnology* **6**.

Oliveira SMA de, Torres TC, Pereira SL da S, Mota OM de L, Carlos MX. 2008. Effect of a dentifrice containing *Aloe vera* on plaque and gingivitis control: A double-blind clinical study in humans. *Journal of applied oral science* **16**: 293–296.

Oncina R, Botıa JM, Del Rıo JA, Ortuño A. 2000. Bioproduction of diosgenin in callus cultures of *Trigonella foenum-graecum* L. *Food chemistry* **70**: 489–492.

Parekh J, Jadeja D, Chanda S. 2005. Efficacy of aqueous and methanol extracts of some medicinal Plants for potential antibacterial activity. *Turkish Journal of Biology* **29**: 203–210.

Pitchai D, Manikkam R, Rajendran SR, Pitchai G. 2010. Database on pharmacophore analysis of active principles, from medicinal plants. *Bioinformation* **5**: 43–45.

Pithayanukul P, Ruenraroengsak P, Bavovada R, Pakmanee N, Suttisri R, Saen-oon S. 2005. Inhibition of *Naja kaouthia* venom activities by plant polyphenols. *Journal of ethnopharmacology* **97**: 527–533.

Ponou BK, Barboni L, Teponno RB, Mbiantcha M, Nguelefack TB, Park H-J, Lee K-T, Tapondjou LA. 2008. Polyhydroxyoleanane-type triterpenoids from *Combretum molle* and their anti-inflammatory activity. *Phytochemistry Letters* **1**: 183–187.

Quisumbing E. 1951. Medicinal plants of the Philippines. *Department of Agriculture and Commerce, Philippine Islands Technical Bulletin.*

Raju J, Gupta D, Rao AR, Yadava PK, Baquer NZ. 2001. *Trigonella foenum graecum* (fenugreek) seed powder improves glucose homeostasis in alloxan diabetic rat tissues by reversing the altered glycolytic, gluconeogenic and lipogenic enzymes. *Molecular and cellular biochemistry* **224**: 45–51.

Reddy KN, Reddy CS. 2008. First red list of medicinal plants of Andhra Pradesh, India-conservation assessment and management planning. *Ethnobotanical Leaflets* **2008**: 12.

Rezaeian S. 2011. Assessment of Diosgenin Production by *Trigonella foenum-graecum* L. *in vitro* Condition. *American journal of plant physiology* **6**.

Rout G, Samantaray S, Das P. 2000. *In vitro* manipulation and propagation of medicinal plants. *Biotechnology advances* **18**: 91–120.

Roy SC, Sarkar A. 1991. *In vitro* regeneration and micropropagation of *Aloe vera* L. *Scientia Horticulturae* **47**: 107–113.

Sarker KP, Islam R, Islam A, Hoque A, Joarder OI. 1996. Plant regeneration of *Rauwolfia serpentina* by organogenesis from callus cultures. *Plant Tissue Cult* **6**: 63–65.

Shahid M, Shahzad A, Malik A, Anis M. 2007. Antibacterial activity of aerial parts as well as *in vitro* raised calli of the medicinal plant *Saraca asoca* (Roxb.) de Wilde. *Canadian journal of microbiology* **53**: 75–81.

Shih MF, Cherng JY. 2012. *Potential applications of Euphorbia hirta in pharmacology*. InTech.

Singh B, Dutt N, Kumar D, Singh S, Mahajan R. 2011. Taxonomy, ethnobotany and antimicrobial activity of *Croton bonplandianum, Euphorbia hirta* and *Phyllanthus fraternus*. *J. Advances Dev. Res* **2**.

Singh GD, Kaiser P, Youssouf MS, Singh S, Khajuria A, Koul A, Bani S, Kapahi BK, Satti NK, Suri KA. 2006. Inhibition of early and late phase allergic reactions by *Euphorbia hirta* L. *Phytotherapy Research* **20**: 316–321.

Singh Sunil Kumar, Yadav RP, Tiwari S, Singh A. 2005. Toxic effect of stem bark and leaf of *Euphorbia hirta* plant against freshwater vector snail *Lymnaea acuminata*. *Chemosphere* **59**: 263–270.

Subbu RR, Chandraprabha A, Sevugaperumal R. 2008. *In vitro* clonal propagation of vulnerable medicinal plant, *Saraca asoca* (Roxb.) De Wilde. *Nat Prod Radiance* **7**: 338–341.

Tapsell LC, Hemphill I, Cobiac L, Sullivan DR, Fenech M, Patch CS, Roodenrys SJ, Keogh J, Clifton PM, Williams P. 2006. Health benefits of herbs and spices: the past, the present, the future. *Faculty of Health and Behavioural Sciences-Papers*.

Tota K, Rayabarapu N, Moosa S, Talla V, Bhyravbhatla B, Rao S. 2013. InDiaMed: A Comprehensive Database of Indian Medicinal plants for Diabetes. *Bioinformation* **9**: 378–380.

Vida F. 1952. Therapy of hypertension with Indian Rauwolfia serpentina. *Die Medizinische* **20**: 1157.

Waman AA, Umesha K, Sathyanarayana BN. 2008. First Report on Callus Induction in Ashoka (Saraca indica): an Important Medicinal Plant. *IV International Symposium on Acclimatization and Establishment of Micropropagated Plants 865.383–386*.

Watson DJ. 1947. Comparative physiological studies on the growth of field crops: I. Variation in net assimilation rate and leaf area between species and varieties, and within and between years. *Annals of Botany* **11**: 41–76.

Witkus ER, Berger CA. 1947. Polyploid mitosis in the normal development of *Mimosa pudica*. *Bulletin of the Torrey Botanical Club*: 279–282.

Youssouf MS, Kaiser P, Tahir M, Singh GD, Singh S, Sharma VK, Satti NK, Haque SE, Johri RK. 2007. Anti-anaphylactic effect of *Euphorbia hirta*. *Fitoterapia* **78**: 535–539.

Zamora AB, Gruezo SS, Siar SV. 1998. Somaclonal variation: breeder's tool for recovering potential cultivars in ornamentals (Kalanchoe). *Philippine Journal of Crop Science* **23**.

YOUR KNOWLEDGE HAS VALUE

- We will publish your bachelor's and
  master's thesis, essays and papers

- Your own eBook and book -
  sold worldwide in all relevant shops

- Earn money with each sale

Upload your text at www.GRIN.com
and publish for free